RUFUS

The Remarkable True Story
of a Tamed Fox

RUFUS

The Remarkable True Story of a Tamed Fox

ERNEST DUDLEY

Hart Publishing Company, Inc., New York City

FIRST AMERICAN PUBLICATION 1972
SBN NO. 8055-1034-6
LIBRARY OF CONGRESS CATALOG CARD NO. 71-186667

MANUFACTURED IN THE UNITED STATES OF AMERICA

Contents

Cast of Characters

THE PEOPLE

Don MacCaskill, a forest ranger
Catherine MacCaskill, his wife
Ian Mackenzie, a young forester
Mrs. Mackenzie, his aunt
Joe Allan, a gamekeeper
Jean Allan, his wife
Meg Allan, their daughter
David Stephen, the vet

THE ANIMALS

Rufus, the fox
Frieda, his vixen
Cassius, the cat
Shuna, the dog
Cronk, the raven

RUFUS

The Remarkable True Story of a Tamed Fox

1

A Life Spared

Two men and their dogs waited outside the fox's den in the forest above Ballachulish. Lights from the little Scottish town on the shore of Loch Leven flickered through the twilight mist. As the forest darkened, the men and their dogs, a collie and a terrier, concentrated on the entrance to the fox's earth.

The man with the collie held a shotgun under his arm. Beside him lay a vixen, most of her head blasted off by shot. The vixen's mate had died earlier that afternoon, caught in the jaws of a steel trap and finished off by the man with the shotgun.

The dog-fox and his vixen had been young. All that

remained of their family now were the young cubs, hidden in the darkness of the den. They had been a first litter—tiny, mewling, furry shapes. When her mate had failed to return with food, the vixen herself had left the den to search for sustenance for her cubs. As she emerged from the den, the men were waiting for her.

The hunter with the terrier finished rolling his cigarette. It was time to finish off the fox family. He put a hand on the dog, and when it gave an excited yelp, he put the animal's head into the den. The terrier disappeared, stumpy tail wagging. From the deep recesses of the den, the dog's muffled yap could be heard. To the cubs huddled together in terror, their dark earth, so warm and secure, had become a death cell.

Soon, the men above heard a worrying gurgle from the terrier's throat, followed by a sudden silence. The dog had made his first kill. Then he reappeared, tail first, dragging out a small twitching, furry body. The man with the cigarette took the dying cub from the terrier, who needed no word to send him back once more into the earth.

The hunter threw the cub down beside its mother. After one final twitch, the bundle of fur lay still. From below, the terrier's muffled yapping turned into a throaty growl again. Silence followed. Then the wagging tail showed once more as another cub was dragged out.

Listening to the terrier scrabble down for a third kill, the man took a deep pull of satisfaction from his cigarette, and watched a thin spiral of smoke vanish quickly on the darkening air. Again, the terrier backed out with a third twitching, furry shape. The collie whimpered while the moon emerged from a straggle of cloud. Back went the terrier. His muffled yelps ascended in pitch.

As the terrier dived down into the earth yet again, the man with the collie exclaimed softly, and with a swift movement dashed forward. When he stepped back, he held a ball of struggling fur in his hands. It was a fifth fox-cub, that had miraculously blundered its way out of the earth through another tunnel, to escape the death-dealing canine. The man stared at the small being in his hands, felt its tiny heartbeat. His partner took the cigarette from his mouth, spat out a shred of tobacco, and reached out to grab the cub.

"Nae, let's spare this yin," said the hunter holding the cub, as he shook his head.

The cigarette's glow reddened, as though to illumine the surprise in the smoker's face.

With his shotgun under one arm, the man holding the cub placed it in the curve of his shoulder. As the collie leaped up, teeth glistening, he cuffed it down with his free hand. "I ken a lad in Ballachulish will give me ten bob for it," he said.

The other hunter gave a shrug. It was twice that, twenty shillings, for an adult fox's tail; only ten shillings per cub before it leaves its earth.

The terrier had turned aside from the entrance to the den; he had made his last kill that night. The dog-fox, the vixen, and the four cubs were shoved into a gamekeeper's bag. The man with the cigarette moved off, the terrier trotting after him, nose lifted to the scent from the bag. The man holding the tiny cub called his collie to heel and followed into the night.

The fox-cub, later to be known as Rufus, had just made his first contact with the world of human beings.

2

Rufus Settles In

Don MacCaskill was a forest ranger in the little Highland village of Inverinan. Early in September of 1968, six months after the hunting episode, he received a telephone call from a Mrs. Mackenzie who lived about fifty miles away in Ballachulish. Mrs. Mackenzie said that her nephew had a young fox he would have to get rid of. She had heard that Mr. MacCaskill was interested in having one.

Yes, Don said, he had wanted to raise a fox. As a matter of fact, he had already prepared an enclosure and a den, hoping that a fox would turn up somehow to take residence.

"Well," said Mrs. Mackenzie, "in that case . . ."

In a moment, everything was settled. Don MacCaskill could fetch his fox.

The next day, about noon, Don and his wife Catherine set off in their Dormobile. Catherine had looked forward to this trip through the Western Highlands and thought they might take time to enjoy the loveliest parts. The road wound through forest and glen, through tiny villages, with glistening lochs on either side. But her husband gave no thought to all this. His mind was so fixed on seeing the fox that he could scarcely keep within a sensible speed limit.

At the very idea of this fox, Don's blood tingled with excitement. Feeling the grin of anticipation on his face, he tried to control it. After all, the fox mightn't be what he expected; it might be too wild, and impossible to tame, or else too docile, with all the life whipped out of it.

They crossed Connel Bridge and had a picnic lunch by the roadside that overlooked shimmering Loch Creran. Catherine pointed out seals basking in the sunshine on the loch's little isles. But Don didn't get out the field glasses. He wasn't interested in seals, or white sea swallows swooping over their nests on the rocky islets, or anything else. He just sat there, hoping that the fox would turn out to be all he wished. He wanted to talk about the possibilities; would it be in good health, or would it be ill, affected by its term of captivity? But his wife had lunch ready and felt hungry.

After lunch, they passed through Appin and the ruined tower of Castle Stalker. Then the slate roofs of Ballachulish came into view, fashioning a bluish glint against the forest beyond. Mrs. Mackenzie and her nephew lived on the outskirts.

When Don rang the bell of the neat, small house, just before five o'clock, he discovered that he had arrived too soon. Mrs. Mackenzie explained that her nephew Ian wasn't back from work in the forest yet. So the MacCaskills took a stroll around the village, Don making poor efforts to conceal his emotions as they returned to the house. Ian still hadn't arrived, but Mrs. Mackenzie assured the visitors that he would be back any minute. Meanwhile, Don politely declined an offer to see the fox while he waited, preferring to wait for young Mackenzie to introduce them. He felt that a fox, like other animals, might not enjoy meeting a human being who was a complete stranger.

A few minutes later, Ian Mackenzie appeared, quiet and pleasant. Don sensed right away that the young man really didn't want to surrender his fox, and was doing so out of sheer necessity. He planned to leave home in a week or so for a job some distance away. Since he couldn't continue to look after the fox himself, nor could his aunt, the only remaining course was to find a home for the animal with someone who promised to take good care of it. He was talking about the fox as if it were a cherished family dog, Catherine thought.

He wasn't selling his pet, though Don offered to pay whatever he had given for it, even a bit more. But Ian Mackenzie didn't want any money, just the surety that his fox would be happy. It was a vixen, he told Don, and well adjusted to people, because he had always taken her with him when he went off to work in the forest. In the truck with his workmates, she sat between his knees. While he worked, she would wait quietly on a collar and leash. He fed her during the day, and when they got home in the evening, she went into her homemade kennel for her evening meal. Then young Mackenzie usually took her for a walk, on which she met the local dogs and became friendly with them. Again, Catherine thought, he was making her sound just like a dog.

While Catherine chatted with Mrs. Mackenzie, Don accompanied Ian to the yard where the fox was kept. He noted the cleanliness of her kennel. When the fox saw the stranger, she seemed a bit reluctant to show herself, but Ian took her up in his arms. She was in good condition. Her well-brushed coat had lost the woolly blue and brown fur of a cub; now she looked rangier, though not yet having acquired the darker muzzle of the fully-grown adult. Don suddenly realized how extraordinarily tame she was.

Young Mackenzie carried the fox to the Dormobile. Don climbed in and tied her by her leash to a protruding leg behind the driver's seat. She was wearing a dog collar, which he silently disapproved. As a precaution, against possible carsickness, he spread a plastic sheet on the floor.

Catherine seated herself, carefully eyeing the fox, who constantly watched Ian Mackenzie. He gave her a farewell pat, then stood by his aunt. Don could see the young man's emotional struggle at parting from this wild creature he had raised from a cub and gentled as a pet. Glancing back while driving off, Don saw Ian Mackenzie turn quickly and go into the house.

Back in Inverinan, someone waited for the MacCaskills —someone almost as eager to see the fox as Don himself had been. This was fourteen-year-old Meg, the only child of their neighborhood friends, Joe and Jean Allan. The night before, Meg had learned from Don that he was going over to Ballachulish to see about getting a fox. In spite of her own eagerness to see the animal, Meg hadn't told her father about Don's errand. She knew that Joe Allan, a gamekeeper renowned for his skill as a stalker of wildlife, had an implacable hatred of foxes. To him, they were vermin that killed pheasants and carried off baby lambs: "Rip up a sheep at lambing time, a fox would, just for its milk." Meg's mother sided with him: "Murderous, cunning things that would eat every egg, slaughter every fowl in a poultry run, in one go." So Meg said nothing to her, either, about Don MacCaskill's trip to Ballachulish.

In the Dormobile, the fox sat gazing at Catherine with a gentle expression, quite unafraid. Don took the eastward road for Glencoe, into the wild and savage glen. Suddenly, the fox stood on her hind legs, her front paws on the back of the driving-seat, as if she wanted to see where she was going. Catherine put out a hand gingerly to pat the fox's head. Her ears went down, but not aggressively, more like someone feeling timid with strangers. The creature's tractability seemed amazing to Catherine.

During the long drive home, the fox jumped up and down with her paws on the driving-seat several times, and Don could hear her slither about on the nylon seatcovers. Occasionally, he glanced back to make sure all was well. Near the end of the trip, the road to Inverinan became narrow and winding, and just as they came within sight of the village, the fox was sick. Don congratulated himself on having put down the plastic sheet.

By the time they got home, it was half past eight, and would be dark soon. Don wanted to give his new friend the run of the enclosure before night fell, so that she would get to know her home right away. He saw Meg Allan by the gate and called out to her as he stepped down from the wheel. Then he went round to the car door as Catherine got out. The fox stood there, looking a little pathetic, and Don picked her up. As he did so, he saw unmistakably that this was no vixen; it was a dog-fox. He wondered how Ian Mackenzie could have made such a mistake, and could only conclude that he must have been misled because the animal had been so gentle. Don held the fox in his arms. He lay there, perfectly calm, and the musky smell of his fur wasn't very strong.

Meg stared, hardly able to speak, but full of wonder at the sight of the fox, tame and gentle, in Don's arms.

"What's its name?" she whispered.

"Rufus," Don replied on the spur of the moment.

Meg repeated the name to herself as she softly stroked the fox's head. Rufus raised his jaw so that she could scratch him. He was obviously used to being stroked and petted. Meg marveled at the roundness of the fox's eyes—not a bit sly, as she had expected them to be, but trusting and innocent.

Catherine hurried into the house to prepare supper, while Don, with Rufus in his arms, went outdoors. Meg followed him, longing to hold the creature, but not daring to ask if she might. Perhaps later on, she thought.

Don's land extended about an acre, and he had enclosed one corner, sixty feet by fifteen feet, with a high wire-netting fence. This area, topped with similar netting, included a small tree. For the newcomer, Don had built a warm, waterproof den, banking it round with earth and small rocks to give it a natural appearance; but the roof could be lifted up for a look inside. The den had two entrances: Don remembered that a fox liked to be able to enter one way and exit another, if it wanted to. As much as possible, the fox's new home really simulated natural conditions.

When Meg opened the gate of the enclosure, Don put Rufus down, removing his collar and leash and hoping that the collar mark would quickly disapppear. Suddenly, Rufus sensed his freedom and leaped into the air with delight. Then he raced from end to end of the enclosure, exploring each corner.

Though it was beginning to get darker, Rufus had enough light to see where he was; and as Don and Meg watched, he paused by the den. Then he burrowed down into a tunnel, coming out again from the other one after a few moments. "It's all right now," Don told Meg. "He knows exactly where he is." It was a relief to know that Rufus would consider the den his home, going into it at night if he wanted to, or sheltering there if it rained. The shore of Loch Awe was only three hundred yards away. The fox had only to stand on top of his den to glimpse the gleaming water; beyond it were the twin peaks of Ben Cruachan. Rufus would have a wonderful view.

While Meg watched him take the measure of his new territory, Don fetched some food—a dish of meat and dog biscuits. He opened the enclosure-gate again and put it down. Immediately, Rufus came over and began to eat, releasing Don from one final anxiety: getting a young animal to eat in strange surroundings can be difficult. "Better leave him now," he said, setting off for the house. He looked back at the girl's shadowy figure still at the gate of the enclosure.

After supper, Don went out again with a flashlight, to check on Rufus. Meg had gone home. It was a warmish night, and dry, so Rufus hadn't gone into his den. Instead, he lay on a wooden bench Don had fixed up for him just beside the den. Often a fox prefers to lie on something above ground level—the branch of a tree, for instance—and Don was glad to see Rufus taking his ease, with his beautiful brush curved round him to cover his face. Peering over it, he glanced at Don, shadowed behind the flashlight.

"Good night, Rufus," he whispered, and as he turned away, Don caught an answering flicker of the bushy tail.

3

Fulfilling a Dream

Like many other Scots, Don MacCaskill had been brought up to believe that foxes should be ruthlessly exterminated. As a boy, he had heard shepherds and gamekeepers discuss the ways in which these predators went about killing lambs and poultry. To the countryman, nothing was too bad for the fox.

But during his first job as a forester near Glasgow, Don had started to wonder whether this universal condemnation was really justified. Late one morning, Don had come upon a group of local sportsmen, indulging in a customary fox-drive. They had already pursued their quarry in a particular part of the forest, and several victims of their guns lay beside them while they tucked into lunch. The lighthearted men were sitting around with their sandwiches and flasks of coffee and whisky. Suddenly, a burly farmer in breeches and leggings, stood up and, still joking, deliberately emptied his gun barrel at one of the dead foxes lying at his feet. "Take that, you bastard."

This scene made a deep, disturbing impression on Don. None of God's creatures, he felt, deserved the contempt he had just witnessed. Determined to learn more about foxes, Don soon learned that evidence of the fox's depredations was far from conclusive. Meanwhile, he began looking for foxes, observing where they made their dens—the rabbit-burrows they would take over. Sometimes he managed to catch sight of cubs at play, but this wasn't often. The great difficulty about trying to study a nocturnal creature was its ability to stay hidden during the day. The idea of learning about foxes by keeping one himself had not yet occurred to Don. At that time, he couldn't have done so, anyway; the hostile reactions of his fox-hating neighbors would have been worse than if they had discovered him harboring a criminal. And even if he had wanted to keep a fox, he lacked the right sort of space for one.

Then Don and his family moved to Loch Aweside Forest in the Highlands, into a house on a fair-sized plot. Now interested in actually raising a fox, Don started to plan an enclosure which would give the creature reasonable freedom, as well as security. He also constructed a hide-away in the forest for use in observing foxes.

These animals were not easy to observe. Don had learned, for example, that he couldn't rely on a fox returning to the same spot where he had last seen him hole up. The next day, when he went back to look for him, the fox would have moved to some other place of cover. But during breeding time, this pattern changed; the vixen stayed in the den with her cubs, while the dog-fox brought them food.

So early in 1968, at the season when vixens give birth to their cubs, Don tried to find a den he could keep under surveillance. Unfortunately, whenever Don tried to foresee which den a vixen might choose, she wouldn't make up her mind. First it would be one rabbit-burrow, then another. Finally, Don was convinced that the vixen he was watching had made up her mind. The den she seemed to have chosen was in a tiny valley, overlooked by a forest ridge. In one of the trees on this ridge, he built his observation post.

Don had noticed that foxes seldom look up, and he was equally aware that those which especially interested him would likely approach their den from below. Therefore, he was not likely to attract their attention. Also, the wind would carry his scent high over their heads. The hideaway itself was built of branches and heather, to be as inconspicuous as possible. All in all, Don felt certain that his presence would remain undetected.

Dealing with foxes, however, is dealing with the unpredictable. The very first time Don got up to his post he heard a bark *behind* him. For some reason, the dog-fox was approaching the den, not from below, but from the ridge above. He had got Don's scent and was barking a warning to his vixen in the den.

Don climbed down and went home in the gathering dusk. It was bad luck, but he didn't give up. The next afternoon he returned to learn the effect of the dog-fox's warning bark. On his earlier visit, Don had believed the vixen was sheltering a litter of cubs in the den. Had she and her mate decamped overnight with their young, or were they still there?

This time Don approached the den from a different direction, and inched forward as quietly as he could. Still hidden by trees, he suddenly heard a bark again. This time it was the vixen, who had picked up his scent. Perhaps she had even heard him—her ears would be as sharp as her nose. Don guessed she was barking to warn the cubs down in the earth, and that meant the cubs were still there—she hadn't shifted them yet.

But she would now, he was certain. Both she and her mate knew that their den had been discovered; instinctively, they would feel they had to move the cubs. Don came back the following day and, sure enough, there was no longer any sign of them. They had stolen away in the night.

Although not surprised, Don felt discouraged. Heightening his sense of defeat was the fact that the pair of foxes he had located were comparatively rare in the district. Loch Aweside was sheep-farming country, preserved by gamekeepers who shot at sight any moving thing that might be a danger to the sheep. Between gamekeepers and shepherds, the prospects for fox survival were grim. It was at this point that Don set to work constructing on his property an enclosure for a fox he could study and come to know. Then one day came the telephone call from Mrs. Mackenzie and the fulfillment of his dream.

The Waiting Game

Don knew that his first objective must be to gain Rufus's complete confidence. He hadn't been deceived by the fox's friendliness on the way home from Ballachulish. Rufus had been on the leash then, under a certain amount of restraint. He couldn't get away, and he knew it. Don regretted the collar and leash as an indication of the way Rufus had been raised, except when shut away for the night in the kennel. It was a miserable life he wouldn't wish on a dog, let alone a wild thing like a fox. He didn't blame Ian Mackenzie who couldn't be expected to know any better. He had kept his pet in good shape physically, at least, but he couldn't read a fox's mind or comprehend how the animal was reacting to continued curtailment of his freedom.

Don recalled how Rufus had sprung into the air when freed of his collar as he was put into the enclosure. For the first time in his life, Rufus had been given a chance to jump and leap about. Every line of his dancing body showed his joyous delight as he raced round the enclosure, reveling in the open space. No longer fettered by the collar round his neck, the tugging leash holding him back, Rufus reverted to a wilder state.

Don realized that he would need to devote much time and patience to Rufus, if they were going to become friends. He must talk quietly with him, appear relaxed, move unhurriedly. He had determined to make no attempt to approach him, and he would not try to call or coax him. He was just going to wait for Rufus to come to him. Certainly Rufus needed time to settle down in his fresh surroundings, to feel secure in his new home. Don felt he should learn everything for himself without anyone keeping an eye on him.

The next morning on his way to work, Don took a look at Rufus, just to make sure he was all right after his journey and that he had food. He seemed slightly restive. That evening, when Don let himself in through the wire gate, Rufus still watched him nervously. He was offered food but didn't want it. Don considered him for a moment, then nodded. "Good night, Rufus," he called softly and left. He didn't even glance back to see if there had been a responsive flick of the brush, but he doubted it.

After that, Don visited Rufus twice a day, always relaxed and unhurried, with a few quiet words. Rufus ate his food, when he was alone; he looked well and alert. But for a week he did no more than glance at Don.

Young Meg Allan came every afternoon to ask after Rufus and to spend some time watching him. She would talk with him very quietly, not attempting to have him come to her. She stood by the enclosure for long minutes without stirring, in a state of sheer enchantment, and Rufus stared back at her with his round, innocent eyes. Then he would turn and lope off gracefully, brush held low to balance him as he looked back at her and moved away again, with a swift movement. After a while, when she quietly called good-bye, he would turn his head to stare at her. Sometimes she had the impression that he was about to make a move toward her. But he remained where he was. She would turn to see him as she went round the corner of the house, and he always watched her out of sight. But for a week he did no more than look at Meg and Don.

Then one day, Don saw Rufus start toward him very cautiously. He kept absolutely still, trying to seem as casual as he could. Very slowly, Rufus came all the way up to him and gave Don the chance to pat him gently on the head. Then he backed away and moved off. At last, patience was going to pay off.

The next evening, when Meg came by, Don had already gone to the enclosure. She caught up with him as he was going in. Rufus stood by his den, apparently absorbed in contemplating several sparrows on the wire-netting roof. Meg was about to say that she would come back another time, because she knew how Don preferred to be by himself while making friends with the fox. But he had her come into the enclosure, telling her how Rufus had come to him the previous evening and permitted himself to be patted on the head. She let herself in, moving slowly, to join Don. The sparrows flew away, and as if their flight now gave him an opportunity to give Meg his undivided attention, Rufus turned toward her. Don remained perfectly still, and Meg had followed his example.

Rufus now concentrated his attention upon Meg, with his head cocked slightly to one side, his round eyes inquiring. It was a long scrutiny. She held her pose, still as a statue. Then the fox's body relaxed. Still staring at Meg, he made a sudden leap into the air, as if pouncing upon something. Don explained to Meg in an undertone that Rufus was pretending to hunt a mouse. Then the fox gave a few more leaps, still without taking his eyes off Meg, and pounced on his imaginary prey. She was careful not to laugh or move. Rufus stopped and carried out a series of leaps, looking at the young girl the whole time. It was almost as if he were inviting her to join him in some sort of game.

Don smiled to himself, but Meg remained poker-faced. She was filled with an indescribable affection. The warmth of her love for Rufus made her long to hold him close, but she continued still. Rufus sat down, evidently puzzled by his failure to make any impression upon her. Don knew that at any moment now the fox would make a move toward her, so with just a word he moved slowly out of the enclosure. He pretended to take an interest in growing things, but kept a watchful eye on Rufus, waiting for him to make a move toward Meg. She acted as though about to follow Don, then turned and waited. Rufus hesitated, then came to her slowly, and allowed her to put her hand on his head.

Meg stayed with him for several minutes, talking to him quietly, feeling the warm softness of his fur. She put her face against his, smelling his warm, musky scent. Don watched them, until it was time for Meg to go home.

When she had gone, he entered the enclosure again. Shadows stole across the land; over in the direction of Loch Awe a solitary star glimmered. About half an hour after Meg had gone, Rufus came over to Don and let him pat his head. Then he jumped away, came back again and lay down, his bushy tail curved in front of his nose, while his friend talked to him. When Don stood up and went quietly out, murmuring, "Good night, Rufus," he sensed rather than saw an answering flick from the brush.

5

Rufus Becomes a Star

Soon word got round the village that there was a tame fox on Don MacCaskill's place. Meg told her school friends all about Rufus and asked permission to bring them to see him. Don was only too pleased. Rufus, although somewhat wary at first, quickly took to the children. The fact of Meg's presence was to him sufficient guarantee of their friendliness.

Meg instructed the children to behave calmly and control their excitement when they got close to the fox. Once they had all made friends, Rufus would run to meet them and stand on his hind legs, forepaws against the enclosure fence, wagging his wonderful tail. Eyes shut in blissful innocence, he allowed the more adventurous of the children to push their fingers through the wire netting and scratch his white chest.

Soon the children were sharing their sweets with him. From a grocer's truck that came round every day and stopped outside Don's house, Meg and her friends bought Snowballs, coconut-and-ice cream concoctions, that proved a great favorite with Rufus. Gradually the children clubbed together to buy him more and more, and Rufus would put away half a dozen at a time with no difficulty at all. What delighted the children most was his way of eating them so that they left a big white mustache across his face—an effect far more bizarre than anything his gratified audience ever managed to achieve.

Don soon realized that Meg's attachment to Rufus had deepened greatly. She spent most of her spare time with the fox, playing games with him or holding him in her arms and talking to him. She found an old tennis ball which she would throw in the air for Rufus to chase. He would allow it to roll still, pretend to turn away, and then, after a show of elaborate nonchalance, pounce on it with a sidelong leap. It was his hunting-the-mouse performance all over again, and he obliged with as many encores as Meg requested.

They also played hide-and-seek, with Rufus diving into his den and Meg guessing which of the two tunnels he would exit. She often tried to catch him as he ran out but never succeeded. He always knew which exit she was waiting at, and would shoot out of the other, twisting and turning at top speed, with his brush flicking to keep him balanced. Meg knew enough not to startle him with unexpected sudden movements when they weren't playing.

She spoke to him quietly and always waited for him to come to her, never chased after him when he was disposed to be on his own.

Don and Catherine had thought of Meg as somewhat shy and reserved, a little younger than her years. She wasn't especially outstanding at her lessons, except in natural history and essay writing. Meg confided to Catherine, however, that she loved writing poetry and that she had written several poems about Rufus, only she hadn't told anyone else about it. Don had a passing acquaintance with Meg's father, a big, handsome man, who wore an open-necked shirt in all weather. He had his daughter's soft voice, even something of her gentle manner. As a gamekeeper, like his father before him, he was reputed to be a crack shot, and none came wiser in the ways of deer and other game than Joe Allan.

Meg had got to know Don when he had visited her school—as he did several times a year—to talk about wildlife and show the children his photographs. In spite of her shyness, she had stood up and asked questions, especially about foxes; what Don said about them had caught her imagination. Later, Meg had called at Don's house, cradling a baby red squirrel with an injured foot, which she had found in the woods. She hadn't liked taking it home, fearing she would be told that the best way to cure the little squirrel was to kill it. Meg had taken this occasion to ask more about foxes. She said that her father hated these animals—always had spoken harshly of them—and she supposed that her own sympathy for foxes had come about because they were so hated by everyone else. It was dif-

Don and Meg took turns
administering the medicine,
which Rufus took
without much fuss.

The only problem
was getting hold of Rufus
when he had to be returned
to his enclosure.
Shuna proved helpful.

ficult for her to believe that God could have created such a "cruel, detestable creature" as the fox was assumed to be, though she never let her father know her thoughts on the subject.

Don then explained the main reason why people round about felt an enmity for foxes: the Highlands was sheep country; and ever since sheep had been introduced a hundred and fifty years before, the fox had been branded a killer. No one took into account the fact that the fox had antedated the sheep by centuries, and that the forest, which had provided him with a staple diet of mice and rabbits, had been cut down to give sheep grazing space. Of course, the fox was a natural predator, and if he couldn't get rabbits, he hunted whatever other food he could find. Lambs were available only a couple of months of the year, and a fully grown sheep was too formidable for a fox to tackle. That left him the remaining ten months in which to provide for himself. Hundreds of sheep died yearly, and these furnished the fox with a certain amount of food. As for Meg's father, well, he was a gamekeeper, and part of his job was to protect the grouse which had been the fox's natural food in the forest. Even though thousands of grouse were bred artificially for sportsmen to shoot, people abominated the fox for wanting a few himself. Indeed, the fox killed many more rats than grouse, and rats were notorious for destroying grouse eggs and chicks.

Stories against the fox had grown into myths, spread because of ignorance and fear. Practically nothing was known of his life history except that he was a nocturnal creature. Human beings still feared the dark; the hours between dusk and dawn were still thought of by some as the time of evil spirits and baleful monsters bent on harm and destruction. The vilest, most murderous cunning was attributed to foxes, simply because they were night creatures. In fact, they were no more artful than other wild animals.

If foxes were all that clever, Don pointed out, why was it that their dens were so easy to find? Why were they not secretly hidden away? One could always locate a fox's earth by the bits and pieces of feather and bone the vixen left outside for her cubs to play with and learn how to hunt. It was really the foxes' lack of cunning and stealth that brought about their destruction; it was precisely because they continued to make a life without instinctively covering up their tracks that they were relatively easy to hunt down.

Meg understood the sense of what Don had been saying. All the same, after the arrival of Rufus, he didn't want her blurting out to her father a lot of notions that must sound like blasphemy to him. He was careful to point out to Meg that with some people, hating foxes was like a religion. It would be like banging one's head against a brick wall to try to make them change their beliefs of a lifetime overnight. You just had to hang on to what you believed to be true, despite all prejudice, and wait for a chance to speak out when you thought this would do the most good. Meg agreed. She feared her father's reactions if he learned that she had such ideas in her head, even though he had smiled indulgently at her interest in Don

It was a curious look of abstraction
on Rufus's face...

...as if his thoughts
were somewhere else
far away.

A fox and a dog together!
Who would ever believe it?

MacCaskill's tame fox.

She would write an essay on Rufus, she told Don, for her next natural-history lesson. In addition to her secret poems, she had already written essays about wildlife at school. Her teacher had even asked her if she had thought of studying to become a veterinary surgeon, but Meg didn't believe she would be able to pass the exams. Anyway, her parents couldn't afford to pay for the training, if she did.

Rufus gradually became a magnet for children from all over. Families driving through Inverinan would stop at the house to ask if it was true that a marvelously tame fox lived there. Some people brought their cameras, and Don was always willing to persuade Rufus to pose for his picture. Unlike many animals, the fox was only too pleased to oblige. Quickly, he would rise on his hind legs, front paws stretched up against the fence to show off his chest and fine tail, his white teeth glistening in a wide grin of joy.

One weekend, the visitors included a family from Edinburgh—father, mother, son and daughter—who had heard about Rufus. The amount of photographic gear unloaded by the father showed how seriously he took his hobby. Rufus posed with customary eagerness, but the photographer, not altogether satisfied, asked if he could take some more photographs from inside the enclosure. The wire netting was a hindrance, he explained. Don decided that he couldn't cause any harm and allowed the man inside.

Don had expected, however, that Rufus might now follow the usual pattern of animals and back away from a stranger. Far from it! He immediately came up and pushed his nose right into the lens as the visitor tried to get him in focus. Instead, it was the photographer who found himself backing away. And then, when at last he was about to click his camera, Rufus dodged behind the photographer and charged off with his exposure meter, which he deposited in his den before dashing back for a further piece of equipment.

Watching him go through such antics, Don suddenly realized that Rufus was now an individual, having developed a personality of his own. He had learned to play games for fun, for the amusement of himself and others. He even seemed to possess a sense of humor. Or, if that was too much to believe, one was forced to admit that the games Rufus played had nothing to do with animal instinct.

Sometimes, to the delight of children crowding round, Don would carry Rufus in his arms like a baby. The tame fox didn't seem to mind how much his brush was pulled or how much he was scratched under his chin, or how much his ears were tweaked. Everyone marveled at his tameness. Sometimes he hardly seemed like a fox at all.

6

A Winter Escapade

Winter arrived, and with it the first snowfall. Rufus slowly came out of his den, contemplating the strange white blanket with wonder. He had never seen snow before. His round eyes narrowed questioningly as he tried to recognize surroundings so drastically changed overnight. Rufus had gone to sleep in a world shaded from gray to black. Now, everywhere, another element appeared: white—sparkling, shimmering white. Feathery flakes fell endlessly, settling at his feet without so much sound as a sigh.

Rufus glanced back at the imprints of his footsteps; as he trotted on, they trotted after him. He jumped up to snap the flakes between his teeth, but they brought only an icy, wet sensation to his tongue. Feeling a whiteness that clung about his face, above his eyes, Rufus shook himself; powdered snow sprayed off his shoulders. For a moment, his face wore a startled expression. Then he decided that there was nothing to worry about. Exhilaration filled his being as he raced round in a shower of white. By now, Rufus had grown into a fine, handsome fox, his biggish bones well covered, the flanks filling out. His brush was magnificently full, and the snow beautifully set off his winter coat.

The snow cleared, and three weeks after Christmas, milder weather arrived. One midday, a passing truck driver pulled up outside Don's office and put his head round the door. On his way to work, he told Don, he had seen a fox.

"Whereabouts was that?"

"About a mile up the road."

The driver gave a description of the fox. He had tried to run the animal down, he said, but it had just trotted on ahead. After unsuccessful attempts to catch up with it, the driver had seen the fox turn into a quarry half a mile away. Still determined to kill it, he had driven into the quarry after it, but the fox had jumped onto a rock and sat and stared at him. He got out of his truck, picked up a lump of rock, and threw it at the fox, but he missed, and the animal had run off into the forest.

The moment the truck driver had left his office Don hurried out to his car and drove quickly home. Something about the description of the fox and its manner made him anxious. As a matter of fact, he had left home that morning without looking in on Rufus as he usually did. As he hurried along, he felt sick with apprehension. Sure enough, the enclosure gate was open, and there was no Rufus inside.

He paused long enough to notice that the gate was propped open with a piece of stick; it had been opened deliberately, either late the previous night or early in the morning. After quickly telling Catherine what had happened, Don drove off to the quarry. But there was no sign of Rufus there. Don went into the forest, calling out for him, without results.

He spent the rest of the day searching, but still no Rufus. Don guessed that Rufus had been badly scared by the truck driver and probably decided to hide during the day. He would be lying hidden in the brushwood, Don thought, not daring to come out for anyone—not even for him.

Don returned home, saddened by the thought that he might never see Rufus again. Meg and several of the children came over early in the evening. Once they had got over the shock of Don's news, they immediately wanted to go out looking for him. Meg personally rushed out to the enclosure to make sure that the fox hadn't somehow returned. She could barely believe the evidence: an empty enclosure, no response to her call, "Rufus . . . Rufus . . ." Sickened by the thought that he might never be found, she hurried back to the house and set off with Don and the children.

The extensive spruce woods and thick bracken in

Shuna took to Rufus
from the moment she arrived;
and he, as readily to her.

Cronk would happily coo
and jabber away,
fluttering her wings
as if asking Don to feed her.
To her, he was still a mother.

the area around the quarry diminished any hopes of finding Rufus. Don felt that Rufus would wait until nightfall before venturing out in search of food, so along with Meg and the children, he continued to comb the area for a mile or more around, until it was suppertime and too dark to see. A sad, silent group headed back to the village.

There was no news of Rufus the next day, or the next. Every day, Don, Meg, and various bands of children would set out on a new search early in the morning before Don went to work; again during lunchtime; and again in the evening. But Rufus might have been spirited away for all the luck they had. Don and Catherine began to believe that Rufus must be dead or else have been so scared by the truck driver that he would never show his face to a human being again. Meg, on the other hand, had recovered some of her lost hope. She refused to believe that Rufus wouldn't come back; each night she prayed for his safety and for his return.

Two weeks after Rufus's escape, Don was down by Loch Awe on the lookout for a migrant osprey which had been reported nearby. Since there weren't more than twenty of these birds in all of the Highlands and the Scottish isles, he wanted to be sure that the bird spotted was really an osprey.

Don had parked his car near a lodge by the forest and made his way quietly to the lochshore. After an hour of patient waiting without sighting the bird, he turned to go home. As he reached his car, he saw something on the lodge gate. It was a brownish-red object—a brush, he knew at once. Somebody must have killed a fox. The lochshore was more than seven miles from Inverinan, but Rufus could have traveled that distance easily.

His heart turning to ice, Don went to the gate and examined the brush—a trophy for which twenty shillings' reward was paid. With some relief he noted that the dead fox must have been a vixen. Yet Don knew that this same fate could easily be in store for Rufus, if in fact he had not already suffered it.

Don didn't mention the dead vixen to Meg that evening when she accompanied him on another vain search. She continued to hope, though the rest of the children had given up Rufus for lost. The next day passed, and still no news; Don, with Meg, had scoured the area, up and down the roadside, searching the forest, looking in every place where Rufus might be expected to hunt for food. Rufus had had no experience of hunting; he wasn't really old enough to take care of himself properly.

Don kept a lookout at every farm he passed, in the event of a complaint of a missing hen which Rufus might have tried for. He heard no such story, however, and concluded that if he were still alive, Rufus must be living off mice and rats. He doubted if he would go after a rabbit, since he had always behaved in a friendly manner toward any of the children's pet rabbits when they had been introduced to him. Rufus might even have treated a rat or mouse as a playmate, if he encountered one in the forest. By the end of the third week, Don had given up all hope, though Meg still felt that her prayers would be answered.

Late in the afternoon that Saturday, the mother of one of the Inverinan children called at the house. She had seen a fox by the roadside, she said, near a farm not far from

From the start,
Shuna was in the enclosure
with Rufus
or around the garden . . .

. . . or bringing him home
to share the hearthrug.

the village. The fox of her description sounded like Rufus, so Don drove to the place immediately.

It was just the sort of place Rufus might have chosen to hide in during the day. The farm had fairly extensive land around it, and the house itself was surrounded by big trees and rhododendron bushes. Don set out to search the area.

He had hardly emerged from the rhododendron bushes when suddenly, there was Rufus, trotting back and forth on the lawn. Don called to him quietly. Rufus stopped in his tracks, and then came straight to Don, his brush wagging. His ears were laid back and his expression was apprehensive, as though he expected somehow to receive a scolding for all the anxiety and trouble he had caused. Just like a dog, Don thought, speaking to him gently. Rufus crouched close, but when Don made a move to pick him up, the fox turned and dodged him. His tail was still wagging, however. It seemed as though he wanted to be picked up and comforted, but was still afraid.

Don followed Rufus to the edge of the rhododendron bushes, talking to him quietly. He kept his voice down, not only to reassure him, but because he had heard the clucking of poultry nearby and feared that the farmer would come out to deal summarily with this fox whom he would assume was about to attack his hens. Again Rufus crouched, and again he ran off when Don reached for him. This time, Don saw the poultry-run around the corner of the house. So did Rufus, who made straight for the hen-house entrance, where the fowls were settling down for the night. The aperture was invitingly open, the farmer hadn't yet closed it, and to Don's consternation Rufus headed straight in.

The noise of wildly beating wings and hysterical cackling seemed deafening to Don. At any moment, someone would surely appear from the farm. Don couldn't get into the henhouse to reach Rufus; it was locked. He didn't know what to do—and then Rufus reappeared, carrying in his jaws an enormous, screeching hen with half its feathers missing. It was obvious that Rufus had no intention of killing the hen. To him, it was just part of some new game, but what farmer would believe that? Don made a dive and caught him by the tail. Rufus opened his mouth to protest this unwarranted interruption of his game, and let the hen go. Still squawking, but unharmed except for having lost half its plumage, it shot back into the henhouse, from which feathers swirled out. Rufus was now in Don's arms, fondly licking his face; there was to be no scolding. His ribs could be felt very plainly, and it was doubtful if he had enjoyed much to eat during the past three weeks.

Don caught the sound of voices raised in inquiry from the direction of the farmhouse. He didn't wait to offer any explanations, but, clutching Rufus to him, hurried to his car. Rufus sat up beside him during the drive, gazing out with interest at the gathering dusk.

7

Rufus, Cassius, and Shuna

As head forester, Don had long done his best to insure that no unauthorized person would set foot in the forest to hunt foxes; but despite his efforts, he was well aware of the farmers, shepherds, or even rabbit poachers, out to pick up twenty shillings' reward for killing foxes. If such men knew that a fox lived in the neighborhood, they would use the most effective means to ferret him out. It was surely someone like this who had deliberately released Rufus, bent on killing him. That he had escaped destruction then was a miracle. Some intuition must have warned him to slip away into the darkness before he was caught. From the point of view of his safety, Rufus was certainly better off in captivity than in the wild.

But Don knew that there was more to be considered than safety alone. One night out in the forest, he heard a dog-fox call. A moment later, a vixen answered him. The vixen's cry turned his thoughts to a nagging problem that had lain at the back of his mind since the return of Rufus. Oh, he had settled down well enough, obviously delighted to be back once more with those who loved him. But Don realized that the time had come for him to return Rufus to the forest. The motive for obtaining him in the first place no longer made sense. Rufus seemed no longer a wild creature to be studied at close quarters; he had become a friend, as well loved as a domesticated dog or cat. But Rufus was in reality a creature of the forest, and Don felt that he owed him the right to live the life for which he had been born.

How could a human being begin to understand what Rufus's real longings were? He seemed happy, it was true. He played games with Meg and the children, he went through his performances, posed for the photographers, reveled in the admiration and amusement he evoked. One could easily imagine that his life was complete. But who knew what feelings haunted him at night when he was alone, listening to the sounds all around him? Perhaps he, too, had heard a vixen's call. What instinctive emotions would he have experienced at that sound, emotions which no human being, no matter how loving, could understand?

Even Meg, who was probably closer to Rufus than Don, perceived what an enormous gulf separated her from him. Don had warned her never to forget that she was dealing with a fox, and that she should not judge him by human standards. Some experiences she could never share with him. For example, she and Rufus saw the same object very differently: to her, it appeared in color; to him, it was all monochromatic, just varying shades of gray. The view of the sunset behind the purpling hills and the evening star, the sky at dawn over the loch—these sights which thrilled Meg meant little or nothing to Rufus. Compared with Meg, Rufus was short-sighted. But he could rely on his powers of scent and hearing to an extent that she could hardly comprehend. Not only did his senses differ from those of human beings, but also his instincts. He lived in a world of his own that no living person could hope to enter—not even Meg or Don, however much they loved him.

These were the thoughts Don faced as he tried to decide what was really best for Rufus. Many times he had discussed the fox's future with Catherine, and had practically made up his mind to release Rufus in the coming autumn. That season was the natural time for a cub to leave his family, when he had learned all he could from his parents and might be pushed out to fend for himself. So it would be appropriate then, in September or October, for Rufus to leave *his* family—Don, Catherine, Meg, and the children—and go out on his own.

But putting into practice what he knew to be right was not going to be easy for Don. He preferred not to think about it at the moment, but rather to make the most of the time he had left with Rufus. He didn't say anything to Meg about what he had decided. Time enough for that later when the time came to release Rufus to the wild.

Meanwhile, the MacCaskill household was destined to get larger, not smaller. Several weeks after Rufus's safe return, Cassius entered the scene. Cassius was a kitten who had holed up under the floorboards of a derelict house on the outskirts of Inverinan. A workman had called at Don's house one evening and told him of this kitten, which he said was the offspring of a wild-cat father and a domestic-cat mother. Wild cats in that part of the Highlands made their homes high up in rocky cairns, sometimes coming down to the village for food when it was scarce in the hills, and Don was aware that the males had been known to mate with female domestic cats.

Don went along to the house and, sure enough, underneath the floorboards lay a kitten. He managed to reach the feline with one hand, but it eluded his grasp, and spat like a wild cat. "A real Cassius Clay, that's what he is," the workman said. Thus Cassius got his name. The kitten had retreated as far as it could go, so Don found a length of wire and pushed it under the floorboards. He intended to hook the wire through the kitten's fur and drag it out bodily.

At that moment, Don made a discovery. Under the same floorboards, only a few feet away, he found four other kittens and their mother. Obviously, the kitten the workman had found belonged to that litter. Don returned it to the others and went back home. The matter didn't end there, however. Next evening, when Don had been home from work about an hour, he heard a mewing in the shed outside the back door. All five kittens were in a dark corner, with no sign of the mother. For some reason she had decided that Don could look after her family while she went off on her own.

The kitten named Cassius had a typical wild-cat appearance; he looked like a little tiger and possessed a tiger's mentality. If his tail was not as short as a wild cat's, that would be remedied later. Don and Catherine found homes for the other four kittens, but kept Cassius who had immediately attached himself to them. The next step was to introduce him to Rufus.

Normally, foxes are highly antagonistic to cats, however young. Nevertheless, Don felt confident that Rufus would be an exception. Cassius, in any event, didn't wait to be introduced. The very morning after his arrival he found his way to Rufus's enclosure and quickly climbed the wire netting to the top. There he sat, staring down at the fox. As Rufus ran up to him, Cassius jumped down to rub his fur against the wire netting. Rufus gave a preliminary sniff. Then he, too, pushed himself along the fence, rubbing against Cassius through the wire netting.

For a week, Don was too cautious to allow the kitten into the enclosure with Rufus, but with the fence between them, Cassius's first move every morning was to run straight to the enclosure. Every day he and Rufus would go through the routine of rubbing themselves against each other through the wire netting. Cassius would climb onto the roof after this, and spend an hour or so observing Rufus down below.

The following week Don and Meg were outdoors with

Rufus, playing his game of pouncing on the old tennis ball, when Catherine called to Don to come into the house. As Don hurried in, Rufus ran after him. Before anyone could stop him, Rufus was in the living room, where Cassius lay on the hearth before the fire. Rufus took in the scene, trotted over to the fireplace, and stretched himself out side by side with Cassius. He behaved like a dog that had come indoors after a tiring run. From that time on, Rufus was often invited—or invited himself—into the house to spend an hour with Cassius on the hearthrug. They might have been the oldest of friends.

Cassius had no fear of any living thing, man or beast, and this would get him into trouble before long. One day, outside the gate, he was attacked by a passing dog. Cassius prepared to give battle, but soon realized that he wasn't going to get away with it. He turned and ran. But before he had reached the first branch of the nearest tree, the dog had leaped up, nipped his tail, and broken off the end of it.

When Cassius didn't come home for two days, Don and Catherine thought he was lost. On the third day he turned up, the end of his tail still bleeding. Fortunately, about this time, Don had made the acquaintance of David Stephen, a noted Scottish naturalist who was especially interested in foxes. Stephen performed a veterinary surgeon's job on what remained of the tail, and it healed. Cassius now sported a very short-ringed tail which, together with his tiger stripes, gave him the appearance of a genuine wild cat, although his white chest and white feet betrayed his maternal origins.

The cat's first action after David Stephen had attended to his tail was to look up Rufus and bring him back to the house, where they again stretched out together before the living-room fire. Cat and fox, curled up together on the hearth, made an odd enough sight, but soon there was to be a third member of the party.

Two years before, while Catherine and Don were on summer holiday, their seventeen-year-old Labrador retriever had died. But for the arrival of Rufus, Don and Catherine would have acquired another dog. Now, a friend on the other side of Inverinan offered them a Labrador bitch puppy. Don was anxious to have it for Catherine's sake. Rufus, he felt, had been pretty well monopolized by himself and Meg; and since Catherine was indoors so much anyhow, she didn't get many opportunities to be with him. Shuna, as the Labrador puppy was named, would take the place of their old Labrador, who had been a fine, home-loving dog. Shuna took to Rufus from the moment she arrived; and he, as readily to her. If anything, he was more friendly toward her than she toward him. For a moment, she seemed a bit frightened by his strange smell, which wasn't a dog smell, but a musky fox's smell. Within minutes, however, they were playing together happily. Rufus, following his practice at Ballachulish, had already made friends with the dogs in the village.

From the start, Shuna was in the enclosure with him or around the garden, or bringing him to the house to share the hearthrug with Cassius. The three of them played together. At first, Don and Meg stood by, looking on to see that the newcomer met no harm, and also that Rufus would not take this opportunity to slip away. But there seemed no danger of this; the only problem was getting hold of Rufus when he was to be returned to his enclosure.

Shuna proved helpful here. Part of a game called for Rufus to lie on his back while Shuna nosed him over. When Don wanted Rufus in the enclosure again, he would instruct Shuna to keep Rufus on his back by holding her paw on his stomach. Don would then come along, pick up Rufus, and carry him off to be safely locked up in his home for the night.

The three animals made a wonderful trio. Rufus, nocturnal creature though he was, had accommodated himself, first with Don and Meg, and now with Shuna and Cassius, to become a daytime animal. He now spent as much time awake during the daylight hours as he would have spent asleep had he been living in the wild.

8

The Arrival of Cronk

Shortly after Shuna joined the family, Don visited that part of the forest near the place where he had built a hideaway for observing foxes. At the time he had first built it, he had noticed that there was a raven's nest on a nearby ledge. Now, as he stood underneath the ledge, Don remembered the raven's nest, and wondered if it was still there. While thinking of it, he heard a rush of wings. A raven flew upward, and something fell—a dark shape tumbling slowly downward. Don ran toward it and found a young raven lying on the ground. The bird appeared uninjured, except for some missing claws that had been pulled off as it tried to grasp branches and rocks during its strange, frightening fall from a sheltered home.

Don tried to put the young bird back in its nest, but the nest was impossible to reach, so he pulled off his sweater and wrapped the raven inside. It looked to be about three weeks old. The bird had an enormous bill, completely out of proportion to its body, and bright, wicked-looking eyes.

Don took the raven home, improvising a cage out of a box. In the bottom of the box, he built a simple nest of sticks bound with wool. After putting the young raven into the nest, he got some food for it. A young bird has to be force-fed to begin with, and normally, it won't take food from anyone but its mother. But Don used a spoon to shovel down the food, and the chick took it without any trouble. For the next three days and nights, he and Catherine fed it every two hours.

The raven grew, squawked, and attracted Cassius who, of course, had to investigate. Shuna and Rufus also came along. Don and Meg watched as Rufus, Cassius and Shuna inspected this black object in its cage. Cassius sniffed at it. There is a strong, rank odor about the crow family—to which the raven belongs—that Cassius didn't like at first and from which he walked away quickly. Rufus appeared interested, so did Shuna. After this, together with Cassius—though he remained the least enthusiastic—they visited the raven's cage daily, as if to note its progress. The raven, in fact, made good progress and grew rapidly. Don had no intention of taming the bird, merely planning to rear it until able to fly, and then releasing it. He named the raven Cronk because of the squawking noise it made.

While the bird was still in the nest, Don mothered it with pieces of meat. But once Cronk had left the nest and started perching on the piece of branch that Don arranged for it, the raven began to feed itself. Don took the bird to a vet to take care of the claws torn out during its fall down the cliff; after an injection, the swelling caused by the injury healed. At the same time, Don discovered that Cronk was, in fact, a female.

By the sixth week, the raven looked ready to fly. She had been flapping her wings while inside the cage, much to the interest of Rufus, Shuna and Cassius, and Don decided that it was time to let her go. But returning her to freedom turned out to be not so simple. Though Don hadn't consciously tried to tame Cronk, his feeding her had somehow given her the idea that he was her mother.

Don took Cronk out of the cage, where she flapped her wings excitedly, and went outdoors. "Good-bye, Cronk," he said, throwing her into the air. Of course, Rufus was watching, with Meg, and Shuna and Cassius, to see what would happen.

Cronk flew upward strongly, but after a few moments dived down and landed on Don's shoulder. He took her in his hands and, once again, threw her into the air. This time she flew around and landed on top of Rufus's enclosure. Rufus hurried into it, climbed up the tree, and examined Cronk more closely. Cronk promptly bent down and stuck her beak through the wire netting to give Rufus a friendly peck, but he ignored it.

By this time, Don had reached Cronk and took her

off the enclosure roof, putting her down outside. She didn't fly away, but strutted around, just like some kind of black parrot, crying, *cronk . . . cronk . . .* Very gingerly, Shuna went up to her and stretched out her nose to give her a sniff. Cronk put her big, curved beak to Shuna's nose and gave her a little nip, quite gently, a warning not to take any liberties. Shuna retreated and sat down to ruminate over this black, strutting creature. Cassius went over to Shuna and sat next to her, as if to reassure the Labrador that this new addition to the family was going to be a friend, provided, of course, that you watched her beak.

Rufus's real introduction to Cronk came when he was out playing with Cassius and Shuna, and the inevitable old tennis ball. Cronk flew down, landed on the fox's back, and tried to tweak one of his ears. Rufus, for all that he appeared so tame, thought this was going a bit too far. After all, he and Cronk hadn't properly met. Promptly, he shook her off his back and chased after her. She made no attempt to fly away, but faced up to him. Rufus moved in fast. Adroitly, he caught her by the neck and gave her a good shaking. Meg, who was watching with Don, caught her breath. And Don suddenly feared that Rufus might revert to type, so to speak, and bite Cronk. But there was no need to worry about Rufus. Cronk survived the episode all right, and from that moment she and Rufus became close friends. But the raven never attempted to peck Rufus again; she knew her place when dealing with so formidable a character as a fox.

Neither did Cronk attempt to peck Don. Catherine

or Meg or the other children were liable to receive a nip from her, if she felt in a particular mood, but Don could rub his face against hers and she would happily coo and jabber away, fluttering her wings as if asking him to feed her. To her, he was still a mother.

The raven is not really a blackbird, though it looks so. Many colors appear in its feathers—blues, greens and purples—and against a light, Cronk's feathers took on an iridescent green hue. She had very dark, large, brightly glittering eyes, and an enormous bill that looked terrifyingly destructive. The bill was designed for picking at carcasses, because ravens are carrion-eaters. Scavenging is their job, and carcasses can be quite tough. Cronk's beak was so strong that, by using its full power, she could crush a person's finger.

At nightfall, Cronk would perch in a window recess outside the back door. The window was near the inside stairs, and every night going up to bed, Don would switch on the light to see this black figure, head tucked under wing, sleeping soundly. Sensing that Don was there, she would raise her head and peer through the window. After a "Goodnight, Cronk," he would be gone upstairs.

That spring when visitors came to see Rufus, Cronk always appeared on the scene, eyes gleaming, missing nothing. She had quickly understood who held the center of attention and seemed determined to capture some of that attention for herself. She deliberately showed off, flying to the top of the enclosure as soon as she saw a lot of people gathered around for a look at Rufus. Then she would dart over and pull at the women's hats, or jump

down and peck at their ankles. For a while, pandemonium reigned until things got straightened out, and Don or Meg succeeded in controlling her.

During all this fuss, Rufus would temporarily retire from his performance—being scratched on his chest, posing for the photographers—and would lie, eyes closed, waiting for Cronk's antics to come to an end. He didn't seem to be jealous of Cronk, or to experience any sense of being deprived of the spotlight. He just seemed happy to let his friend share in the fun. Only two or three times did Cronk take advantage of the fact that she was protected by the wire netting and give Rufus a nip in his tail. He would look annoyed at this and turn as if to bite her, but it was all really a game.

Cronk soon learned to imitate Shuna. She would perch on the roof and bark away down the chimney, a sound that reverberated throughout the house. Whenever buzzards down by the loch flew over the house, Cronk would chase them away by barking at them like a dog.

One of Rufus's friends was a terrier who used to yap a lot outside the house—the same terrier that had chased Cassius up the tree and nipped off his tail. Cronk would deliberately tease this terrier by barknig at it like another dog. Rufus, Shuna and Cassius—the cat safely ensconced in a tree—would watch while Cronk, perched on the garden fence, barked at the terrier in the road. The dog would go berserk with fury and try to scramble up the fence to get at the barking bird. Cronk would then hop down onto the road, and the terrier would race after her. As soon as the dog got close, the raven would spread her wings and flutter above him, just out of reach. From Cronk's point of view it was just a game, but not so with the terrier. He was out for blood. Luckily for Cronk, she was too fast for him.

By now, Don and Meg were taking Shuna, Cassius and Rufus into the forest. On the first outing, Don had thought it wise to put Rufus on a leash, but after only a quarter of a mile he had taken it off as obviously unnecessary. Cronk insisted on going along, too. But keeping up with the others presented a certain problem for her. If she walked, as they did, she would soon find herself left way behind. If she flew, on the other hand, she would soon be way ahead of everyone else. She solved this problem very ingeniously by riding either on Shuna's back or on Rufus's.

A typical "walk" in the woods for Cronk would start with her balancing herself on Shuna's back, legs outstretched, holding on grimly. Then, after a little while, she would take off and fly ahead, waiting for the rest of the party to come along. Next she would perch on Rufus's back, holding on tightly as she had done with Shuna. She never attempted to ride on Cassius's back, but would trot along beside him with the peculiar sideways hop she had. Growing tired of that, she would hop onto Shuna, who would trot on, tail wagging, with the black apparition balancing on her back, wings outstretched. Then she would take off to land once more on Rufus's back and go through the same performance. Now and again she would fly ahead, looking back at Don, impatient for him to hurry

forward and try to catch up with her again.

Cronk was forever picking up nails and bright pieces of stone on these trips through the woods. She would conceal her findings in the crevices of rocks or in holes of tree trunks, after which she would take moss from the rocks and cover the things that she had hidden. Every time she went back with the others to a familiar forest path, she would fly ahead to find and examine the treasures she had hidden away. Fence staples—galvanized and bright in color—were one of her favorite items; she seemed really addicted to them.

By nature, Rufus was too gentle to take advantage of Shuna's easy-going disposition when they played games. Cassius was sometimes quite rough with her—she always took it in good part—and Cronk constantly teased her with his nips and barks. But even Rufus couldn't resist acting as Cronk's accomplice in a wonderful game she had devised. Shuna would always bring out a bone, just given her, as if to show it off, whereupon Cronk would stalk her; at the crucial moment, Rufus would distract her attention and Cronk would dive in, snatch the bone in her beak, and fly off with it. Just a few yards away, she would drop it and wait for Shuna to dash up and retrieve it. Cronk would permit her to have a few gnaws, then stalk her again. Rufus would distract her attention once more at the crucial moment; Shuna would drop the bone, and again Cronk would "steal" it.

Naturally, Cassius did not wish to be left out of the game. When Shuna became wise to Rufus's part in the plot, and refused to be distracted by him, Cassius would dash in to take her by surprise and, as before, Cronk would grab the bone in his beak and be off with it. Of course, Shuna was always permitted to enjoy her bone finally; after a while, in fact, the game reached the point where she deliberately allowed herself to be the "victim" of the plot. To Don and Meg and the children, the four animals were like a troop of seasoned comedians.

Cronk also extended her activities among the villagers themselves. She had already caused Catherine a certain amount of annoyance by following her around while she was hanging the wash on the clothesline and removing the clothespins so that all the laundry fell to the ground. Then she began visiting some of the gardens of the village. One neighbor, Mrs. McInnes, was very fond of Cronk, but began to complain about the way she dealt with *her* washing, especially since Cronk had singled out Mrs. McInnes as a special friend, flying over to see her every morning!

Cronk also teased the village children when they were on their way to school, wearing their tam-o'-shanters with pom-poms on them. He would swoop down from a chimney or rooftop, snatch a tammy, and fly away with it. The owner would think he had lost his hat forever, and the younger children would be very upset. Meg always reassured them that Cronk would soon return the tammy safely. She always did.

One morning on his way to work, Don noticed a raven high above the village. It wasn't Cronk, he knew; she was with Rufus, Shuna and Cassius. He watched the raven fly around for two or three minutes, then saw it go off in the direction of the forest. Later, in the forest, he saw it again, flying high in the blue sky. Looking back on the incident later, he recalled a stab of intuition that the appearance of the raven had some significance.

Next morning, he saw the bird again, circling high over the house. This time, there was a sudden rush of wings, and he saw Cronk fly up to meet the other raven. They flew together for a time, circling high above the village, calling to each other and diving. Then the strange raven headed for the forest again. Cronk returned to the house. Later that day, when he was in the forest, Don saw the strange bird once more. Even as he watched it, another raven joined it. Cronk. There was no doubt about it.

For several minutes he watched the two birds circling and diving. At the back of his mind, on his return to the office, was a sense of unease. When he got back to the house at lunchtime, there was no sign of Cronk. That afternoon he kept a lookout for her and the other raven, but they didn't appear again. Yet when he returned home that evening, there was Cronk perched on the garden gate, waiting for him.

He gave a sigh of relief and bent his head to hers. She chattered to him, rubbing her black feathery cheek against his face. Soon she was on his arm, fluttering her wings at him in a familiar way as he walked toward the house. When he opened the door and called out to Catherine that Cronk was back, she flew up to the roof. But there still remained that niggling unease; he hadn't erased from his mind the picture of her and the other raven.

9

Rufus Captures the Highland Show

For three days every year, toward the end of June, the Highland Show gains wide attention in Edinburgh. This show is an international display of agricultural items, both livestock and equipment. In 1969, when Don MacCaskill was asked by the Royal Scottish Forestry Society to put on a wildlife exhibit, he decided that Rufus would be the best attraction he could present, especially for children.

Don had to be in Edinburgh the day before the show opened, so that he could make sure Rufus's enclosure had been properly prepared. Accordingly, he set off the evening before in the Dormobile, with Rufus and Shuna beside him on the driver's seat. He had no idea of putting

Shuna on display with Rufus. He had taken her along because he was afraid she might get into mischief at home, where Catherine was too busy to keep an eye on her all the time. Besides, in the unfamiliar surroundings, Shuna would be company for Rufus.

Meg and the other children had come to see Rufus off. Some of them would be going to the show with their parents, but not Meg. Her parents couldn't afford the trip. Edinburgh was 150 miles away but it might as well have been the moon so far as Meg was concerned. She had never traveled beyond her village.

Rufus's enclosure was twelve feet square, securely covered by wire netting. It had a shelter for him to sleep in; straw, food and water were waiting. Rufus settled in right away. Apparently, he could do so anywhere, provided he was with a person he knew and trusted. Right beside the enclosure, Don put up the sleeping tent he had brought for himself. He would not take any more risks than he could help. After all, it was an agricultural show, attended largely by rural people—farmers, shepherds, gamekeepers and the like—to whom the sight of a fox could bring on an attack of such apoplectic hatred that someone might start looking for a gun. Also, somebody might turn Rufus loose again during the night with the object of destroying him.

When the show opened next day, crowds immediately thronged around Rufus's enclosure. Don couldn't leave Shuna tied up on her own, so he put her in with Rufus. The result caused quite a stir. A fox and a dog, supposed to be mortal enemies, playing games together as friendly as could be—the news spread so fast that soon it was impossible to get near the enclosure. Hundreds of people, mostly children dragging at their parents, clapping their hands and calling out, were thrilled by the extraordinary sight. A fox and a dog together! Who would ever believe it?

The children scratched Rufus's chest through the wire netting, and Rufus and Shuna played games while they oohed and aahed and applauded. Don allowed only news photographers and television cameramen inside the enclosure. Rufus enjoyed himself best when there was an opportunity to steal a piece of camera equipment and dash off with it to the shelter, much to the onlookers' delight. Whenever the newsmen wanted to photograph Rufus and Shuna together, one of them always managed to be somewhere else at the crucial moment when the camera clicked. Tame animals are usually harder to photograph than those in the wild, and Rufus and Shuna made matters as awkward as they could.

Rufus enjoyed the first afternoon with the BBC television camera crew best of all. Before anyone knew what was happening, he had his head in a bag, pulling out camera lenses—always his favorite—and anything else he could lay his teeth on, and whipping them off to his shelter. The crowd loved this. The less enchanted, and confused, television men finally managed to get control of the situation. Order was restored, and Rufus, assisted by Shuna, put all he knew into his performance before the camera.

Don and Catherine
found homes
for the other four kittens,
but kept Cassius.

Cassius had no fear of any living thing, man or beast;
and this would get him into trouble before long.

That same evening, Rufus appeared on a news program. Don wasn't able to see it, but he telephoned Catherine, who got in touch with Meg—her parents didn't have television—and invited her over to the house. The young girl was so overcome at seeing Rufus on television that when the program ended she found herself crying. It had been a hard disappointment for her not to be at the show with him.

There were photographs and news stories about Rufus in the Scottish newspapers that evening and the next day. His appearance had been a great triumph. All the publicity drew thousands more to the show, and attendance figures were higher than ever. Unquestionably, Rufus remained the big attraction.

But Don noticed, two or three times during the day, that Rufus would become tired and withdraw from the crowd to rest in his shelter. He was rather like a stage star retiring to his dressing room between acts. Shuna, though much more placid, would follow suit; the two animals would lie down together and take a breather.

Don observed something else that bothered him though he couldn't tell why. It was a curious look of abstraction on Rufus's face, as if his thoughts were somewhere else far away. No doubt, Don decided, being so close to such a crowd was proving a strain. He, too, felt rather tense from having to be constantly on the alert lest some fervid fox hater try to injure Rufus or some clumsy member of the throng take a poke at him with his walking stick.

Most of the visitors were friendly and admiring, but

not all. One woman had wanted to buy Rufus—almost demanded him—because, as she explained, the district where she hunted was temporarily short of foxes. A shepherd barged up to Don and upbraided him for daring to keep a tame fox, when it ought to have been destroyed at birth. If it weren't for people like Don, he said, "there wouldn't be such bluidy vermin all over the place, killing sheep." "I hadn't realized," Don replied archly, "that there were so many people trying to protect foxes, but if there are, I'm glad to hear it."

When Friday came, and it was time to pack up and go home, Don was relieved. Episodes like this had left a nasty taste in his mouth. Deep down, he couldn't help feeling that somehow he had allowed Rufus to be exploited, when he really ought to be doing what was best for him. Once again, he knew that he should be planning to return him to the wild where he belonged.

It was ten o'clock and growing dark when the Dormobile got back to Inverinan. Catherine, Cassius and Meg were waiting at the gate. Meg took Rufus in her arms. He nuzzled his face against hers, and she smelled the wonderful musky scent of him. Followed by Cassius, purring a welcome and arching his back against his legs, Don and his wife went into the house. Meg took Rufus to the enclosure, put him inside, and carefully padlocked the gate. He stared up at her through the wire netting, wagging his tail. She had a chokiness in her throat now that he was back. "Good night, Rufus." He twitched his brush and went slowly over to his den.

Cronk put her big, curved beak to Shuna's nose and gave her a little nip, quite gently, a warning not to take any liberties.

Shuna retreated
and sat down to ruminate
over this black,
strutting creature.

Meg returned to the house with the padlock key. When she came in, Don told her what Catherine had just told him: Cronk was missing. She had flown off in the afternoon and not returned. Don remembered the other raven and that intuition he had had.

10

An Invalid Recovers

Meg thought that perhaps Cronk had left because she was jealous at Don's going off to Edinburgh with Rufus and Shuna, and leaving her behind. But Don didn't believe that. He knew it was inevitable that she should go, and in a way he rejoiced, it was natural for her to find a mate and fly away. He could share her happiness and be grateful for the wonderful memories she had left behind. For several days after Cronk had vanished, Don kept a lookout for her and the other raven; but so far as he knew, he never saw them again, never again heard Cronk's raucous cry in the sky. And the inexorable call she had answered only served to underscore for him the freedom he owed

Rufus. The decision he would have to make was fast approaching.

A couple of weeks after their return from the Edinburgh triumph, Rufus started to go off his food and began to appear ill. His eyes were dull, his attitude listless; he looked thin and out of condition. Had he picked up an infection during the show, or had someone managed to poison him? He wasn't vomiting; if it was poison, he must have got rid of it. But he was certainly shivering. His nose and eyes were running. Perhaps he had suffered a chill or some sort of influenza. It seemed essential to keep him warm, so he was brought into the house and bedded down in a dog basket, padded with blankets. Rufus lay in the basket on the hearthrug, safe from drafts, wrapped in a plaid shawl, and feeling very sorry for himself. He was sufficiently house-trained to stagger out of his basket to a piece of newspaper in the corner. Then he would make his way very slowly and mournfully back to his basket, waiting for his shawl to be wrapped around him again. Covered by the shawl, he would lower himself almost out of sight, with only the tip of his nose showing.

Shuna and Cassius were very sympathetic. They nuzzled him and licked his face and did their best to revive his spirits, but there was no improvement. After three days, Don decided to get help from David Stephen, the Glasgow vet. Stephen thought it might be a case of poisoning or some infection picked up at Edinburgh, but he said he would know better if he came down and examined Rufus himself.

The next afternoon he arrived, took a careful look at

Rufus, and diagnosed what is known as German, or Continental, distemper. Stephen was a firm believer in herbal remedies for sick animals, feeling that they instinctively know which herbs or berries possess the appropriate curative properties. A cat or dog, for example, eats a special kind of grass when in need of an emetic. In the wild, Rufus would have found a cure for himself. Stephen duly prescribed some tincture of belladonna, a bismuth compound and pepsin, and cascara—a teaspoonful of the medicine to be administered four times daily. He felt confident that the patient would make a complete recovery.

Don and Meg took turns administering the medicine, which Rufus downed without too much fuss. Shuna mothered him while Cassius lent his moral support. The two of them nuzzled him and licked his face to express their sympathy, willing him, as it were, to get better. Almost from the first spoonful, he began to perk up, his eyes brightened, and he wanted to eat.

The village children had called every day to inquire after Rufus. Now that the news was better, they brought his favorite ice-cream concoction, Snowballs; but he wasn't allowed to eat them yet. By the end of the week, Don was able to report to David Stephen that his medicine had done the trick.

Stephen had also prescribed egg beaten into amontillado sherry, and this Rufus swallowed with relish—raw egg and sherry being the one thing about his invalid routine that he did enjoy, together with the attention he received. Wrapped in his shawl and huddled in his basket, he looked the part to perfection, perking up when his egg

Rufus lay in the basket
on the hearthrug,
safe from drafts,
wrapped in a plaid shawl,
and feeling very sorry
for himself.

The next afternoon he arrived,
took a careful look at Rufus,
and diagnosed . . .
Stephen was a firm believer
in herbal remedies.

and sherry was due and responding happily to the loving care heaped upon him.

By the beginning of August, Rufus was well enough to give up his basket on the hearth, and was taken back to his enclosure. He expressed great joy at returning to his old home by leaping into the air in his special way and racing round from corner to corner exploring his territory; he dived into the den, reappeared from the other exit, dived back again, and came out the way he'd first gone in. Meg produced the old tennis ball, and he began playing his usual pouncing game. Shuna and Cassius joined in, and the local children came along, too. Then the Snowball man came by, and soon Rufus was wearing again the comical white mustache that delighted the children so.

It was just like old times, except that he didn't settle down quite the same as before, and Don still saw him wear that strange, abstracted look. As the summer ended, it seemed to Don that Rufus had withdrawn still more. Finally, Don made up his mind to pretend to Meg that Rufus had once more escaped. He simply couldn't find it in his heart to tell her he meant to set him free deliberately. It would be less painful for her to think that Rufus had gone of his own accord, like Cronk.

11

The Return to the Forest

By the end of September, Don decided to release Rufus in a place where he would have the best chance of finding food easily. He planned to put out food for him at the release point for the first few days, so that he wouldn't starve while getting his bearings. To begin with, he wouldn't be used to hunting, but he would learn quickly this instinctive trait, and besides, it was the time of year when the young cub in the wild leaves its parents and goes out into the world on his own. Perhaps the season explained Rufus's feeling of abstraction.

Very early one dawn, Don left the house and went to the enclosure. Rufus came over to him as he ordinarily did, chattering a welcome. Don opened the gate, picked him up, and took him to the car.

Loch Awe was reflecting a sky beginning to grow pink as Don reached the shore. Behind him in the forest were the last gray remains of the night. When they came to the release point, Don could hardly bring himself to take Rufus out of the car. He nearly changed his mind about the whole idea, deciding to take him back. He held the animal tightly a few more moments, then put him down. Rufus simply stood there looking at him, making no attempt to move. Don made as if to pick him up again, but this time the fox moved away a few steps. Then he looked back at Don, as if asking him to follow him. Don stood and watched. Rufus trotted along the shore and paused once more. Again he looked back, waiting for Don to come after him. He behaved as if this were a new game he was expected to play. Don stood still as Rufus took a drink from the loch while the water lapped his paws.

Then, gathering his courage, Don got into the car and looked back at Rufus staring out across the loch as it glittered in the morning light. There were little splashes of silver as he began to trot along the edge of the water. Don blew his nose hard and drove off. He had left food for Rufus by the side of the loch, so that he would be all right that day. Then he would come back that evening to leave more—dried meat and rabbit—in the same place.

It was still very early in the morning. Don went back to bed but hardly slept; Catherine, who had awakened, didn't get much sleep either. About six o'clock he rose to make a pot of tea. Catherine had at last fallen asleep, and he went downstairs as quietly as he could. Just as he put the kettle on, he heard a knocking at the back door. Opening it, he found Meg standing there with Rufus in her arms. "He's escaped," she said.

Don stared open-mouthed and closed the door after her, unable to say anything. Still clutching Rufus in her arms, Meg came into the kitchen. While he was licking her face, she explained how she had got up that morning as usual and opened the back door to get the milk. There sat Rufus on the doorstep. He had jumped into her arms and she, in astonishment, nearly dropped the bottle of milk. Without telling her parents, she had hurried over with him through the early morning mist. Now she must get back or her parents would be wondering what had happened to her.

Don held Rufus, while the fox licked his face and chattered to him in his inimitable way. Catherine had heard their voices, and came downstairs just as Meg left. She could hardly believe her eyes when she saw Rufus. Don just let the tears run down his face.

12

Rufus Takes a Wife

Don headed back to Inverinan from Airdrie with a crated vixen in the back of the Dormobile. David Stephen had recently acquired her from the Royal Scottish Society for the Prevention of Cruelty to Animals; she was a year old and had grown too big for her original owner.

Three weeks had passed since Rufus's return, and he had settled down again after his brief adventure, though his attitude toward Don seemed a little wary at first. Don now realized that in trying to put Rufus back in the wild, he had misunderstood his needs. Rufus had grown too tame for such an abrupt change. His attachment to his human friends had become unbreakable; he was caught between the human world and the wild. But Don knew that he still had a responsibility to provide Rufus with a bridge linking the two. The answer was to find Rufus a mate. Don turned to Stephen with his problem, and the result now lay in the back of the Dormobile.

Her name was Frieda, and she seemed entirely different from Rufus. She was smaller in size and not nearly as friendly as Rufus had been. She had almost cowered when Don first met her. Stephen himself had been able to handle her, but, he explained, she had liked his two dogs.

Don was halfway home when suddenly, out of the corner of his eye, he saw Frieda sitting beside him. He promptly braked. The door of the crate stood open, and Don thought of putting her back inside but hesitated. One can never be on the same terms with an animal once it's taken a bite of you, and he knew that in trying to move her he would run that risk. He drove on as Frieda continued to sit there. After a couple of miles, she jumped down and quietly returned to the crate. She had decided that it was her home.

Don reached Inverinan after dark. Since he couldn't foresee how Rufus would react to Frieda at their first meeting—or she to him—he decided not to put her in the enclosure right away. Instead, she could spend the night in the shed at the back of the house, and introductions could be postponed until the next morning. He prepared a corner for her with plenty of straw, food and water; let her out of the crate, which he left for her on the chance she might prefer it; and carefully locked the shed door behind him.

Next morning, she was out of the crate and he couldn't get her back into it in order to transport her to the enclosure. She crouched in a corner, raising her lip at him and snarling. If he attempted to pick her up and carry her bodily, she would struggle, bite, and escape. She had decided the corner of the shed was her abode, and there she wanted to remain. Finally, after several minutes of talking to her quietly and shifting the crate into another position, he persuaded her to return to it.

Rufus caught her scent the moment they approached the enclosure and was at the gate as Don went in. Don put the case down and opened the door. Frieda came out at once. For one heart-stopping moment, Don thought she was going to attack Rufus. But Rufus put up his nose to hers in his usual, friendly way, and soon they were both wagging their tails. Don sat on the crate and watched them. Rufus followed Frieda around, sniffing at her like a dog at a bitch. She seemed quite happy to be with him and they sat down together, licking each other's faces.

At the beginning of his relationship with Frieda, Rufus paid somewhat less attention to Don than usual, but his affection toward Meg never wavered. Frieda, however, always kept herself very much to herself. Once she had sorted out the territorial bounds of the enclosure and the den, her attitude was remote. Even Rufus in his most abstracted mood seemed a hundred times more friendly than she. Meg and the children were very excited at watching the vixen with Rufus, but Frieda soon took herself off, usually to the den, where she remained most of the day; Don observed that she was more active above ground at night.

Not for her the Snowballs that Rufus gobbled up. Whereas he still stood up against the fence with evident bliss to allow his chest to be scratched, still posed for photographs, and went through his performances with the old tennis ball, Frieda seemed anxious not to be noticed; and, as Meg at once observed, she didn't possess the same round, innocent eyes that were such a marked feature of Rufus.

But Don knew that Rufus was happy. Frieda represented the link he needed between his life with human beings and the life he might have in the wild. He still enjoyed the games and companionship with Shuna and Cassius, however, and came often to the house to share the hearth with them. He still played with Meg and the children, as well as with passing tourists who came to see him in increasing numbers.

When such visitors appeared, Frieda invariably remained down in the den. The presence of people didn't seem to distress or upset her; she just didn't show any interest in them; and Don was satisfied that she suffered no harm as a result of Rufus's popularity. She lived her own life.

Frieda ate at the same time as Rufus, and kept up communication with him. He made an understanding mate, never intruding on her solitariness. Just as he had solved the problem of leading a nocturnal existence on the one hand and sharing daytime hours with Don and Meg and the others, so now he took part appropriately in Frieda's life. This was what struck Don as Rufus's most extraordinary and endearing trait—his ability to divide his life among all those who mattered to him.

Toward Shuna, Frieda seemed reasonably friendly. She was a female, and anyway, Frieda had been friendly with David Stephen's dogs. But it wasn't the same with Cassius, who took one sniff at Frieda from a safe distance through the wire netting, turned his back, and went his way. Frieda seemed to return his sentiments.

As winter approached, Rufus warmed again toward Don; it was as if he had learned to appreciate what had been done for him and was determined to express his love as fully as he could. The snow fell very heavily that year, and this gave him marvelous opportunities for playing games with Frieda. The two of them would go racing round and round the enclosure, sending the snow flying. When Don came to watch, Rufus would deliberately race up to him and stop short, covering him with a snowy shower. Don would reciprocate by chasing Rufus and piling snow on him. They played in the drifts together, and it didn't matter if Frieda refused to join in; Rufus was demonstrating his love for Don.

In February, Frieda came into heat. She started barking and calling at night, after the manner of her kind. It was a strange, screeching call, which Don recognized from the forest. Foxes breed once a year, and Don hoped that Frieda and Rufus would mate, but he knew that he mustn't count on it. First of all, Frieda was only a year old and still seemed rather unsettled in her environment. Besides, it was rare for foxes to mate in captivity. But, once again, Rufus proved himself a special case. Don witnessed Rufus and Frieda mating in mid-February of 1970. Would she give birth to cubs? Would he at last have the opportunity to observe the way a pair of fox parents raises its young?

Don was sure that Frieda had reached the first stages of pregnancy when he saw her pulling the fur away from her teats one day. He thought she was beginning to look thinner around the flanks; then he realized that in fact her belly was becoming larger. Her head seemed smaller on her body, and she began appearing even less frequently in the enclosure, spending more and more time in the den. At the sight of anyone, even Don or Meg, she would slink off underground. Finally the time came when she stopped leaving the den altogether.

Don made no attempt to disturb her or even to try to look at her; he kept away. The danger was that if she were disturbed when she gave birth to her cubs, she might destroy them. Knowing that the gestation period of a fox is about fifty-two days, it was possible to work out the date when the cubs would arrive. It seemed safest to wait at least a fortnight after the due date before risking a look at them.

Meanwhile, Don's job was to provide Rufus with food for Frieda. Whenever Meg or the children came to see Rufus, he was always brought outside, and Meg would see that everyone dropped his voice so as not to disturb Frieda. Shuna and Cassius, too, weren't allowed to spend as much time in the enclosure as before, though Rufus came into the house with them frequently.

One day, Don told Meg and the children that according to his calculation, Frieda must have just given birth to her cubs. From that day on, the children were absolutely on tiptoes with suspense. Just about two weeks after this date, Don first saw Frieda come up for the food Rufus had put out for her. A lot more fur had been pulled away from her teats, and there was the moisture of milk on them. The cubs were there!

They would be tiny black things, only a few inches long; born blind, their eyes would not open for fourteen or fifteen days. Frieda could be expected to suckle them underground until they were a month old. Then she would allow them out of the den. As that time approached, Rufus began calling them. It wasn't the same ecstatic chattering with which he greeted Don and Meg, but a low-pitched wuffling. It was meant to get the cubs to come out and take a look at the world about them.

One morning as he was about to leave for the office, Don heard Rufus calling more loudly, more insistently, more enticingly than ever before. Instead of leaving, Don went over to the enclosure. Sure enough, the cubs were just emerging into daylight for the first time—two of them. He didn't go into the enclosure but remained outside watching.

Frieda appeared, keeping her eye on the cubs and seeing that Rufus didn't get too near them. Don saw her curl her lip at her mate when he did make a move toward them. He couldn't help feeling a certain chokiness in his throat as he saw the two young cubs.

Their coats had already changed to a smoky color. He could see clearly their opened eyes, which were an absolutely brilliant blue. Both of them saw him immediately as he stood there and stared intently, not a muscle of their tiny bodies moving. At this age their faces were not at all fox-shaped—they didn't yet have the long fox nose—but looked somewhat squat. They hardly resembled foxes at all. A fox litter usually numbers three or four—even five. Don therefore concluded that perhaps because Frieda was so young, she had given birth to only two.

Don was very pleased at the cubs' arrival. It would compensate for the failure of his observation post in the forest. He didn't intend that the cubs be tamed any more than Frieda had been; he wanted them to grow as freely and naturally as they could. This was important for the field study he planned. Watching the cubs in the enclosure would offer him clues by which he could guess with reasonable accuracy the real conditions of foxes living in the wild.

In their natural state, as Don had observed in the past, the cubs would be fed in the morning or in the evening, depending upon when the dog-fox killed and supplied food to the vixen. Don knew how much food the cubs needed. He prepared that amount and added it to what he fed Rufus and Frieda every evening. It was the forest which supplied the cubs' menu: dead rabbits, dead birds, rats and mice. He fed Rufus and Frieda dried meat, but he wouldn't try that on the cubs at this age, because dried meat was quite hard and their teeth were not yet fully developed.

Don found later that Frieda had burrowed into the floor of the den. He hadn't wire-netted the floor when he had built it, so the earth was bare. Frieda had tunneled her way into it for nearly ten feet. She never allowed Rufus to go into this freshly made tunnel, keeping him out with a raised lip and a snarl.

When the cubs came up from below ground, however, Frieda began giving up to Rufus some of the responsibility for their care. In fact, as Don pointed out to Meg, she retreated by degrees into the background until, by August, she was having almost nothing to do with the cubs. Most of the time she would lie down in a corner, or else she would climb the tree and stretch out on its topmost branch where the cubs couldn't reach her. If one of them did get near her, of course, it would attempt to suckle her. By now the cubs' teeth were very sharp and extremely painful for Frieda.

So Rufus virtually took over her job. He gave less time to entertaining the children and spent hours playing games with the cubs: hide-and-seek, chasing each other, and, of course, the pouncing game with Meg's old tennis ball. Don and Meg noticed that he was always careful not to let a cub stray too far from the den; whenever he saw one of them wander off, he would always catch it and bring it back. He was as loving and gentle as any father could be.

As summer turned the corner into autumn, Don noticed that the cubs—one of which was a dog-fox and the other a vixen—were beginning to go off individually to discover what lay beyond the area round their den. Rufus was now less inclined to fetch them back. They were becoming old enough to fend for themselves; soon they would want to go off on their own, to live their own lives.

That October, Don was offered the job of head forester at Tummel Forest, Pitlochry, Perthshire. He would be leaving Inverinan the following month.

13

Farewells and Au Revoirs

If the move to Pitlochry meant a momentous upheaval for Don and Catherine, what did it mean to Meg? For her, it spelled good-bye to Rufus; she would be fortunate if she ever saw him again. Pitlochry was sixty miles of narrow, winding roads from Inverinan, and half a day away—remote, isolated, with no rail and few buses. Anyway, the fare was impossibly expensive for her, even if her father or mother would allow her to go. Don was so dismayed at the thought of having to break the news to Meg that his wife offered to do it for him. But that evening, when Meg came around, he told her.

At first, she couldn't really comprehend that it meant

the end of her friendship with Rufus, who had become so much a part of her life. Don and Catherine tried to reassure her by suggesting that she try to persuade her father to let her visit Rufus. But she was certain he wouldn't allow it, and as for his giving her the fare for the journey—this was something she couldn't bring herself to ask. She would be talking about something he was incapable of understanding: his hatred for the "pestilent vermin" hadn't diminished just because he'd been indulgent about her making friends with "MacCaskill's bluidy fox."

Next evening, however, Meg had wonderful news. She had mustered enough courage to speak to her father, told him Rufus was going, and pleaded to visit him—at least later, perhaps after Christmas. She had begged with all her heart—and he had relented. She could go. He would give her the bus fare. Meg then went to see Rufus and held him close. He chattered to her, licking her happy tears away, and appeared to understand that everything was going to be all right after all.

Don wasn't taking the cubs. He had discussed their disposition with David Stephen, and together they had decided to release them, giving them their chance in the wild. Since Don already had had a chance to study their family life and take pictures of them, scrupulously avoiding any attempt to tame them, now was the time to grant them freedom. Their new home would be a wildlife reserve near Loch Lomond. Stephen would come over to help catch the young cubs and transport them to the release point.

Don confided their plan to Catherine, but to no one

else; and one evening, two weeks before the MacCaskills' departure for Pitlochry, Stephen arrived. The two men went into the enclosure. At their approach both cubs dived into their den. Freida proved no problem, for she showed no interest in them and went off into a corner, while Rufus was persuaded to keep out of the way.

This left Don and Stephen to dig the cubs out—the only way to get them, since they were instinctively scared of human beings. When the men had dug down to the end of Frieda's burrow where the cubs lay huddled together, they grabbed them. Then they put the cubs into a big crate Don had made, placed the crate in the Dormobile—along with some rabbit and dried meat—and drove off.

Two hours later the men reached the wildlife reserve on the shore of the loch. There they set the crate on the ground and waited for the pair to show themselves. But the cubs didn't want to come out: even after such a short journey, the crate had become their home. It was quite a job to persuade them. Don and Stephen had to shoo them, bang on the crate, even tip it a little. Finally the young dog-fox poked his head out very cautiously and looked around. Then he grew bolder and came right out, sniffing with his nose in the air.

It was about seven o'clock in the evening, with dusk falling. Owls hooted and waves were lapping the shore. The dog-fox trotted along the shore a little way, then came back and invited his sister to appear. She came out and went up to him. They were rangy-looking, with eyes the same shape as their mother's—not round and innocent like Rufus's—but with dark fur like his.

The cubs turned away from the shore together, jumped onto a bank, and disappeared into the trees. Don had put down their food before they left the crate; they had actually gone by it, but they knew it was there.

Finally, Don and Stephen took the crate and drove home. Every night for a week, Don brought food to the release point, but the last evening found it uneaten. The cubs had learned to hunt for themselves successfully. They were now self-sustaining in the wild. Perhaps they would stay together, even mate, and have cubs of their own. Don liked to think they would and that they would elude their enemies.

The evening arrived when Meg knew she held Rufus in her arms the last time for months to come. She would be at school next morning when the Dormobile left; the other children already had said their good-byes to Don and Catherine, Rufus, Frieda, Shuna and Cassius on their way home from school that afternoon. Meg wanted to say good-bye on her own, and Don left her alone with Rufus in the enclosure. Both Don and Catherine were careful to look at Meg very casually when she came back to the house. But she wasn't crying, there were no tear-stains on her face, and when Don went with her to the garden gate, she still held back her tears. He watched her disappear into the evening dusk, walking slowly, and he stood looking after her for several moments after she had passed out of sight.

The road from Pitlochry, a little town surrounded by

sheltering hills, branches off over Garry Bridge and winds toward the forest. Lying beside it was the MacCaskill's new home, a low-built white house, with a large garden in front. Behind it stood offices and outbuildings. Down from the back ran a lightly wooded slope, and it was here that Don had built an enclosure for Rufus and Frieda. He had planned and set it up while he was waiting to move in, so it was all ready when he and Catherine arrived. This one measured four times the size of the enclosure at Inverinan, and was strongly wire-netted, top and sides. Don had dug a splendid, most natural-looking den. Other appointments included a bench for Rufus and several small trees for Frieda to climb.

Since the location was very popular with tourists, especially in the summer, Don realized there would be crowds wanting to watch Rufus; but the experience at Edinburgh had made him wary of putting him on show again. For another thing, he couldn't be sure what sort of reaction there would be among local people when it became known that he kept Rufus and Frieda. He couldn't rule out the chance that attempts would be made to kill them both, for there were just as many people round Pitlochry who implacably hated foxes as there had been at Inverinan. According, he had planned the enclosure to take advantage of a protective terrain; by trees and the rise of land, it was screened from the road the tourists used.

On the crisp, clear days of winter, the icy mountaintops of Glencoe were visible far away in the west. The snows passed and Rufus seemed happy in his new home, with plenty of space to move around. From his bench he could enjoy the loch and the forest below. His old friends, Don, Shuna and Cassius, were on hand, and soon he was sharing again the living-room hearth; Frieda remained as aloof as before. The foxes would mate in the spring, and Don would be able to continue his studies of the fox's life history.

Early in February of 1971, Meg came home from school one afternoon to be met at the door by a neighbor. The doctor was there also to tell her that her father had died suddenly of a heart attack and her mother was in a state of collapse. Though it took Meg several weeks to get over the shock of that dreadful afternoon, she was able to write a brief letter to Don and Catherine, who already knew of Joe Allan's death. She asked after Rufus, sending him her love, but told them that she had to look after her mother now. Already ailing, Mrs. Allan was depressively ill as a result of her husband's death, and this meant that Meg's leaving home for any length of time would be impossible.

She wrote every now and again, asking after Rufus, but there seemed little chance of a visit in the foreseeable future. By the end of May, however, Don felt that Jean Allan ought to allow Meg a few hours off, and decided to try a little persuasion. Setting off on the narrow, winding road to Inverinan one day, he visited Mrs. Allan, who agreed that her daughter could go back with him for one afternoon.

Meg was full of excitement, although worried that after so long a time Rufus wouldn't remember her. She was greeted by Shuna, who wagged her tail with delight at seeing an old friend again, while Cassius purred away, arching his back against her. Catherine said hello, and then Don took Meg down to the enclosure. Rufus, standing by the gate, looked up at their approach. Suddenly, he caught Meg's scent and momentarily froze. Then he jumped up against the fence, chattering with ecstasy. Meg could barely speak to him she was so choked. Then she went in. Rufus, jumping up and down with joy, leaped into her arms. As he nuzzled her and licked her face, she caught his familiar musky smell.

All afternoon, Meg remained with Rufus. She had brought the old tennis ball, which had been hidden away with her notebook full of poems, and they played his old pouncing game. Frieda hadn't shown herself at all, but stayed in a corner down the slope. At teatime, Rufus came into the house with Meg and sat next to her, while Shuna and Cassius joined the party.

It was getting late; Meg had promised to be home before darkness. She and Rufus returned to the enclosure while Don waited for her in the car. After a short while, she brought back the key to the enclosure gate, which she had carefully locked after her. Then she got into the car and Don drove her back to Inverinan. She said very little on the way home, and he didn't encourage her to talk. But as they approached the village, where lights were beginning to show in the windows, he said, "You'll come again . . ." and she looked up at him with a quick, hopeful smile.

List of Illustrations

ST. PHILIP'S COLLEGE
00031355 H